Cause and Effect Is a Lie

Also by The Quantum Shifter

The Truth of Reality
The Biggest Misconceptions About Manifestation
Why Isn't It Here Yet?
Manifest In Steps
Identity Shifting
One Decision Creates a New Reality
Self-Concept Selects Your Reality
It's Just a Reflection
Manifest Faster
Identity Shifting
I AM PERFECT
Embody a Wealth Identity
Mental Diet Mastery
Stop Spiraling While Manifesting
Emotional Mastery
Tell Yourself This When You're Depressed
Manifesting Money 101
Money From Nothing
You Are The Bank
Money Mindset For Single Moms
Your SP Loves You
Manifest A Celebrity
Have The Best Sex of Your Life
Lose Weight The Mental Way
Manifest Away "ADHD"
The Writer Everyone Reads
You Don't Need 3D Proof
Action Is Optional

The Positive Delusion
Cause and Effect Is a Lie
Objective Reality Is a Myth
Revision
Collapse & Command

CAUSE AND EFFECT IS A LIE

Why Assumption Creates Reality (and Nothing Else Does)

The Quantum Shifter

IDENTITY PRESS

Connect with The Quantum Shifter

Website: www.TheQuantumShifter.com

Facebook: www.Facebook.com/TheQuantumShifter

ISBN: 978-1-972432-25-9

Published by Identity Press

These works were written over many years during the author's ongoing study of identity and reality creation.

They are now being published for the first time under the pen name *The Quantum Shifter*.

YOUR REALITY IS WHAT YOU SAY IT IS.

Contents

Before You Continue

This book will make the most sense if you are open to a simple idea: Your reality is not fixed. It is not happening to you. It is reflecting what you assume is true.

If you believe that reality is solid, unchangeable, and determined by circumstances, parts of this book may feel confusing or difficult to accept. And that's okay. You don't need to force yourself to believe anything. You can simply set this book down and return to it later, if and when it feels right.

But if something in you already senses that reality may not be what it seems—if you've noticed patterns that don't quite make logical sense—or if you've experienced moments where things shifted in ways you couldn't explain— then this book will likely feel familiar.

Not because it's teaching you something new, but because it's putting words to something you already know. Read it slowly. Let it land. And most importantly: Don't try to "do" anything. Just see what you notice.

There is only assumption
and reflection.

Introduction: The Most Taboo Truth

Cause and effect is the most unquestioned law in our culture. From the time you're a child, you're told that everything happens for a reason. If you study, you'll get good grades. If you work hard, you'll get ahead. If you act right, you'll be rewarded.

This story runs so deep that most people don't even realize it's a story. It feels like absolute truth. Science, religion, and even manifestation teachers frame reality as one long chain of cause and effect.

But here's the taboo truth: there are no causes and no effects.

There is only assumption and reflection.

Why This Feels Radical

When you hear "cause and effect is a lie," your first instinct might be to resist. It feels obvious that events cause outcomes. It feels natural to believe that action produces results. But look closely, and the cracks appear.

Two people can take the exact same action and get completely opposite outcomes. Two people can experience the same event and spiral into opposite directions. If the action or the event was the true cause, the results would be consistent.

They're not. And that's because action and events aren't causes. Assumption is.

The Real Sequence

The real loop of reality looks like this:

1. You assume something.
2. Reality reflects it.
3. You interpret that reflection as "evidence" of a cause.

But the evidence came second. The assumption came first.

That's why your life doesn't change when you frantically adjust actions or try to fix external events. **It changes when you shift the story you're assuming.**

My Turning Point

For me, the turning point came when I realized I had taken the same actions in business, relationships, and money — yet sometimes they "worked" and sometimes they didn't. The difference wasn't effort. The difference was what I assumed about myself in that moment.

When I assumed failure, even the best actions fell flat. When I assumed success, even the smallest moves exploded into results.

That's when I understood: the chain of cause and effect had always been an illusion.

What This Book Will Do

This book will show you:

- How deeply the cause/effect story has been programmed into you.
- How to spot when you're reacting to events instead of creating from assumption.
- How to sever the illusion of linearity with revision.
- How to collapse the middle and stop

needing "steps."

- How to live from the only real law: assumption → reflection.

Once you see that there are no causes, only stories, you are free. Free from the past, because nothing it contains can dictate your future. Free from struggle, because you no longer need a sequence of steps to justify your success. Free from doubt, because you know the only law is the one you declare.

This book is not about working harder, thinking longer, or stacking up rituals. It's about unraveling the most dangerous assumption of all: that cause and effect exists.

When you drop that illusion, life simplifies to the only loop that ever mattered: **assumption and reflection.**

“When I change the story,
I change the outcome.”

Chapter 1
The Myth of Linearity

From the time we're young, we're taught to see life as a straight line.

Do the homework, get the grade. Work hard, get the promotion. Be a good person, receive good things.

It's a comforting story because it promises predictability. If you take the right steps, you'll end up where you want to go. But anyone who's lived long enough knows it doesn't actually work that way.

Why Linearity Feels True

Linearity feels true because it matches how we're conditioned to think. Culture, science, and even religion reinforce the idea that life is one long chain of cause and effect.

- Science: action produces reaction.
- Religion: sin produces punishment, obe-

dience produces reward.

- Culture: hustle produces success.

It all boils down to the same assumption: if you do X, you'll get Y.

But reality doesn't behave in a straight line.

The Cracks in the Story

Look closely and you'll see the contradictions:

- Two people study equally hard. One gets an A, the other struggles.
- Two entrepreneurs hustle day and night. One breaks through, the other burns out.
- Two people treat their partners with devotion. One marriage thrives, the other collapses.

If linear cause and effect were the law, the outcomes would always match the inputs. But they don't.

What actually determines outcomes isn't the action — it's the assumption underneath the action.

The Real Driver

The student who assumes "*I'm smart, I succeed at this*" stabilizes success, even if their study methods aren't perfect. The entrepreneur who assumes "*I'm inevitable, my work is valuable*" stabilizes breakthroughs, even if their strategies are messy. The partner who assumes "*I am chosen, I am loved*" stabilizes devotion, even if arguments arise.

The cause isn't the step. ***The cause is the story.***

The Comfort and the Trap

The myth of linearity is seductive because it seems to give you control. Do the right steps, and you'll get the right results.

But the trap is this: when the results don't show up, you assume you didn't do enough. You double down on action, spiral into struggle, and reinforce the belief that life is hard.

All the while, the truth was simpler: your assumption created the result before the action even began.

Identity Statements

- "Life is not linear. Assumption decides, not steps."
- "Actions don't cause results; assumptions do."
- "Linearity is a myth. Reflection is the law."
- "When I change the story, I change the outcome."

Exercise: Spotting Linearity

1. Think of one area where you've believed "if I do X, I'll get Y." Example: "If I work harder, I'll finally make more money."
2. Write down the evidence where this hasn't held true. Example: "I worked harder last year and didn't earn more."
3. Ask yourself: "What assumption was I carrying that actually shaped the result?"
4. Rewrite the story as assumption → reflection. Example: "Money flows be-

cause I assume it does — not because of how hard I work."

Closing Reminder

The world teaches you to think in straight lines. But reality is not linear. Actions don't produce results. Assumptions do.

When you break free of the myth of linearity, you stop chasing steps and start claiming authority.

Integration Notes

"Nothing outside me has the power to dictate my future."

Chapter 2

Why Events Don't Cause Anything

If cause and effect were real, the same event would always produce the same outcome. But it doesn't.

The truth is simple: events don't cause anyth ing.Your **assumption about the event** shapes what follows.

The Same Event, Two Realities

Job Loss One person assumes: "*This is proof I'm worthless.*" They spiral into fear, avoid opportunities, and sink deeper.

Another assumes: "*This is the push I needed to rise higher.*" They interpret every interview as confirmation of success and quickly land something better.

The job loss wasn't the cause. The assumption was.

Breakup One person assumes: "*I'm unlovable.*" They close off, distrust others, and repeat the pattern.

Another assumes: "*I am chosen and loved.*" They see the breakup as a stepping stone, meet someone new, and experience deeper connection.

The breakup wasn't the cause. The assumption was.

Money Bill One person assumes: "*I can't get ahead.*" Every bill feels like punishment.

Another assumes: "*Money always comes back multiplied.*" Every bill becomes proof of circulation — and unexpected inflows confirm it.

The bill wasn't the cause. The assumption was.

Why Events Feel So Powerful

Events seem like causes because they're visible. You can point to them, replay them, and say, "*This is why my life is the way it is.*"

But look deeper. Two people can share the same event and end up with opposite outcomes. That alone proves the event isn't the driver.

Events are neutral. They only become causes when you hand them that authority.

The Spiral of Reaction

Most people live by reaction. Something happens, they assign it meaning, and that meaning dictates what comes next.

- Event: a rejection.
- Reaction: "This means I'll never succeed."
- Result: doubt, hesitation, self-sabotage.

They blame the event. But the event didn't cause the spiral — the assumption did.

The Power of Creation

When you step out of reaction, you enter creation.

- Event: a rejection.
- Assumption: "This is proof that something better is mine."
- Result: renewed confidence, aligned opportunities, eventual success.

The event didn't cause the success. The assumption did.

Identity Statements

- "Events don't cause outcomes. Assumptions do."
- "Every event is neutral until I assign meaning."
- "I decide the story, and the story decides the result."
- "Nothing outside me has the power to dictate my future."

Exercise: Rewrite an Event

1. Think of one event in your past that you've blamed for where you are now. Example: "I failed that exam, and that ruined my career."
2. Ask: "What assumption did I attach to it at the time?" Example: "I'm not smart enough."
3. Rewrite the event with a new assumption. Example: "That exam was proof I was being redirected to something bet-

ter."

4. Decide: this was always the meaning. Watch how your present begins to shift when the past is rewritten.

Closing Reminder

Events are not causes. They are mirrors waiting for your script.

What follows isn't determined by what happened. It's determined by what you assumed about what happened.

Integration Notes

You declare first, and
evidence follows.

Chapter 3

The Illusion of "Evidence"

The strongest argument people make for cause and effect is evidence. They say, "*Look – this happened, and then that happened. Clearly one caused the other.*"

But what they're really pointing to is not evidence of cause. It's evidence of assumption reflected back.

Why Evidence Feels Convincing

Evidence feels solid because it shows up after the fact. You assume something, reality mirrors it, and then you point to the mirror as "proof."

- Assume: "*I always get rejected.*"
- Reflection: you notice every rejection.
- Evidence: "See? It keeps happening. That's the cause."

But the reflection wasn't proof of cause. It was proof of the loop.

Evidence Is Always Late

Evidence can only ever appear after you've made an assumption. It lags behind the story you stabilized.

That's why you'll never find evidence before belief — only after. And that's also why waiting for proof is the slowest way to create change.

Reality doesn't produce evidence first. **You declare first, and evidence follows.**

Example: Money

One person assumes: "*Money is scarce.*" Every bill, every expense, every dip in the account looks like evidence.

Another person assumes: "*Money always finds me.*" Refunds, gifts, unexpected payments appear as evidence.

Both point to their lives as proof. Both are convinced they're right. And they are — because evidence always matches the assumption.

The Trap of "Proving It"

Many people never stabilize new stories because they keep checking for proof.

- "I'll believe I'm wealthy once I see more money."
- "I'll believe I'm loved once they call me back."
- "I'll believe I'm successful once I get the promotion."

This is backwards. Reality is always late. Evidence doesn't create truth. Your declaration does.

Why "Cause" Is Just a Label for Evidence

When people say "*this caused that*," what they're really saying is, "*I assumed this story, and then I saw evidence.*"

But instead of admitting the assumption, they give the credit to the event. This keeps them trapped in cause/effect thinking, waiting for events to change before they allow themselves to shift.

Identity Statements

- "Evidence doesn't create truth. My assumption does."
- "Evidence is always late – reflection, not cause."
- "Waiting for proof is waiting forever."
- "My declaration is the only cause. Reflection is inevitable."

Exercise: Catch the "Because"

1. For the next day, notice every time you or someone else says "*because*." Example: "I'm upset because he ignored me."
2. Ask: "What assumption is really behind this?" Example: "I assumed I'm not important."
3. Rewrite the assumption: "I am always chosen and valued."
4. Watch how the reflection shifts when you change the "because."

Closing Reminder

Evidence is not cause. Evidence is reflection. And it always comes after assumption, never before.

Integration Notes

You declare what's true,
and events rearrange to
reflect it.

Chapter 4
REACTION VS. CREATION

The fastest way to see whether you're living under the illusion of cause and effect is to notice whether you're reacting or creating.

Most people live by reaction. Something happens, and they let that event dictate their state. They believe the event caused the outcome. But the truth is, it's always their assumption about the event that sets the loop in motion.

Creation is the opposite. Creation means you don't wait for events to tell you what's true. You declare what's true, and events rearrange to reflect it.

How Reaction Works

Reaction follows this pattern:

- **Event:** Something happens.
- **Meaning:** You assign it a negative label.

- **Spiral:** Attention, evidence, and reinforcement strengthen the story.
- **Result:** Reality mirrors the spiral.

Example:

- Event: A friend cancels plans.
- Reaction: "*They don't value me.*"
- Spiral: You remember past times people canceled, feel rejected, and expect distance.
- Result: The friendship cools.

The event didn't cause the outcome. The reaction did.

How Creation Works

Creation flips the sequence:

- **Declaration:** You decide your truth first.
- **Event:** Whatever happens is filtered through that truth.
- **Evidence:** You notice only what matches your declaration.
- **Result:** Reality reflects your story.

Example:

- Declaration: "I *am always valued.*"
- Event: A friend cancels plans.
- Meaning: "*They must trust our bond enough to reschedule easily.*"
- Result: The relationship deepens because you interpret it as secure.

Same event. Different outcome. The only difference is whether you reacted or created.

Real-Time Shift

The moment you catch yourself reacting, you can pivot into creation.

- Pause.
- Ask: "*Am I reacting to the event or declaring my story?*"
- If you're reacting, rewrite the meaning immediately.

Example:

- Reaction: "*That bill means I'm broke.*"
- Creation: "*That bill is proof I'm wealthy*

and money flows easily."

Why This Matters

When you live in reaction, you're at the mercy of events. You're always explaining your life with "*because.*"

When you live in creation, you cut off the illusion of cause and effect. You don't need reasons. You only need decisions.

Identity Statements

- "I don't react to events. I declare my truth."
- "My declaration comes first. Reflection comes second."
- "Reaction is illusion. Creation is law."
- "I never wait for events to tell me who I am."

Exercise: Reaction Journal

1. For one day, write down three times you reacted to an event. Example: "The client canceled, I felt rejected."

2. For each, rewrite the event from creation: Example: "The client canceled, which proves I attract better ones."
3. Repeat this daily until the pivot becomes automatic.

Closing Reminder

The difference between reaction and creation is the difference between illusion and authority.

Reaction says, "*This happened, so that must be true.*" Creation says, "*This is true, so everything that happens confirms it.*"

One keeps you trapped in cause and effect. The other proves it was never real.

Integration Notes

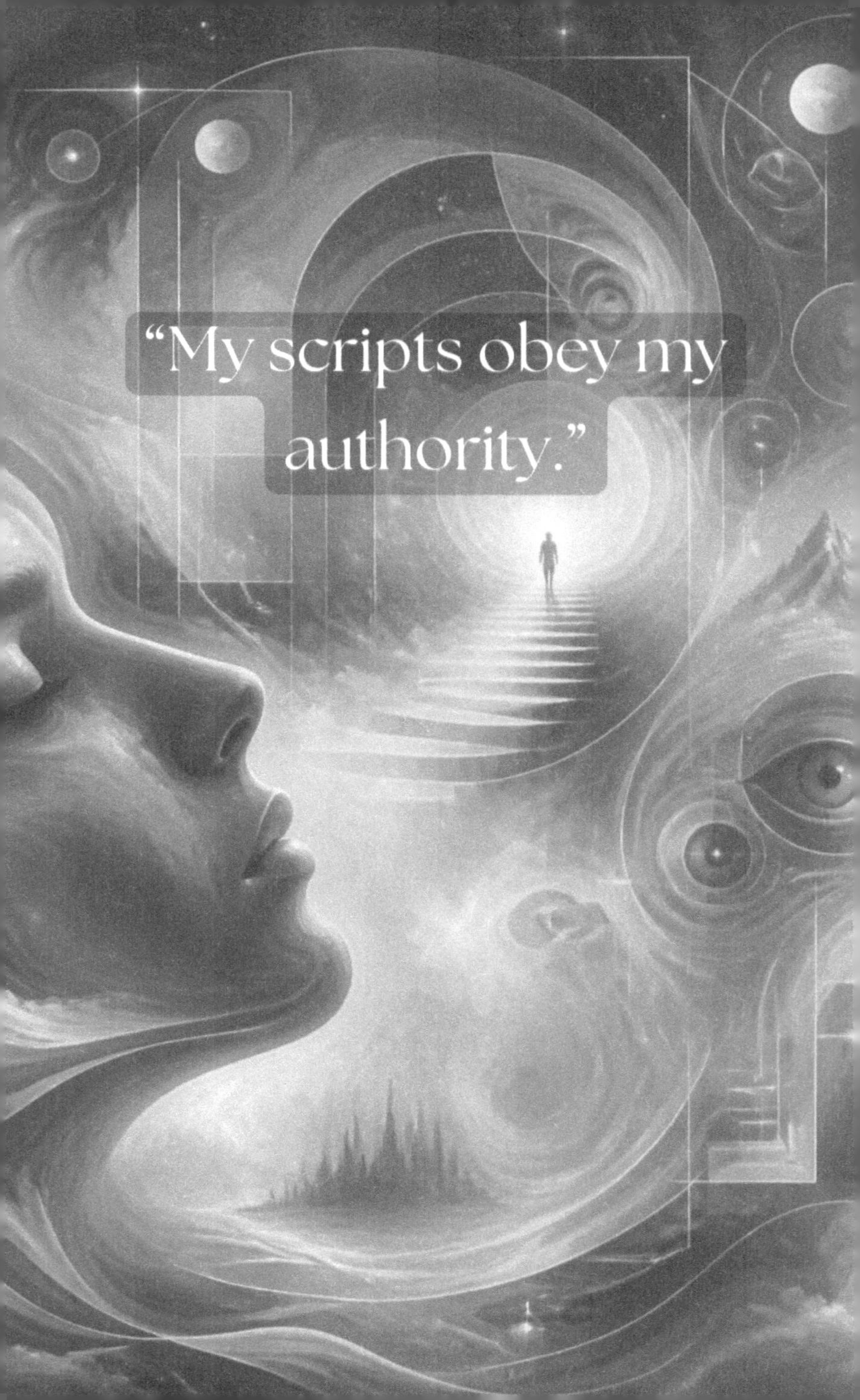
"My scripts obey my authority."

Chapter 5

Observing Your Scripts

Cause and effect doesn't just live in the big moments of life. It hides in the small, unconscious scripts you run every day.

These scripts are the quiet sentences that begin with "*because...*" or "*this means...*" They seem harmless, but they reveal exactly where you've been handing authority to events instead of to your assumptions.

How Scripts Run on Autopilot

Examples of common scripts:

- "I'm tired **because** I worked late."
- "She's upset **because** of what I said."
- "I can't get ahead **because** of the economy."
- "I'm confident **because** I lost weight."

Each one assigns cause to an event. Each one reinforces the illusion that something outside of you dictates your reality.

But what's really happening is this: you assumed the story first, and then you filtered the event through that story.

Spotting the "Because"

The word "*because*" is often a red flag. It's the mind's way of stitching together events into a cause-and-effect chain.

- "I'm happy **because** I got good news."
- "I'm stressed **because** of the deadline."

But the truth is, your happiness or stress came from the assumption you attached to the event — not the event itself.

Observing Without Spiraling

The first step to breaking these scripts is observation. You don't have to fight them. Just notice when you hear yourself assigning cause.

Instead of, "*I'm upset because of this*," pause and ask: "*What assumption did I just attach?*"

This separates the neutral event from the meaning you gave it.

Example: Business

- Old script: "*My launch failed because I didn't market enough.*"
- Deeper truth: The assumption "*It's hard to succeed*" shaped the outcome.
- New script: "*My launches always succeed because I decide they do.*"

The marketing was never the cause. The assumption was.

Why This Matters

Once you spot these scripts, you realize how much of your life is run on autopilot cause/effect thinking.

Every "because" is an invitation to rewrite the story. And every rewrite strengthens the truth that assumption is the only cause.

Identity Statements

- "Every 'because' reveals a script I can rewrite."
- "Events don't explain my life; assumptions do."
- "I catch myself assigning cause and replace it with choice."
- "My scripts obey my authority."

Exercise: Script Audit

1. For the next 24 hours, notice every time you use "*because*" or "*this means...*" Example: "I'm frustrated because no one called me back."
2. Write down the sentence.
3. Cross out the cause. Replace it with an assumption you choose. Example: "I'm always wanted. The calls will chase me down."
4. Repeat until rewriting becomes natural.

Closing Reminder

Your scripts reveal your assumptions. The "because" is never the truth – it's just the story you've been looping.

When you observe and rewrite those scripts, you cut the illusion of cause and effect at its root.

Integration Notes

"I rewrite memory, and reality obeys."

Chapter 6

Severing the Illusion with Revision

One of the most powerful ways to break the spell of cause and effect is through revision. Revision severs the false chain between "this happened" and "so that must follow." It cuts the thread of linearity and replaces it with authority.

Why Revision Works

Cause and effect depends on memory. You look back and say: "*This happened, and because of that, this is my reality.*"

Revision interrupts that story. Instead of treating the past as fixed, you rewrite it as proof of your chosen identity. The past becomes just another reflection of assumption — one you are free to change.

Breaking the Chain

- Old loop: "*I failed the exam, so I can't succeed in this field.*"
- Revision: "*That exam was proof that I rise higher. It was the turning point where success began.*"
- Old loop: "*The breakup happened because I wasn't good enough.*"
- Revision: "*That breakup was proof that I am chosen, and it cleared the way for deeper love.*"
- Old loop: "*I got sick because I pushed too hard.*"
- Revision: "*That illness was proof my body is powerful and recalibrates into greater strength.*"

The event didn't cause the outcome. Your assumption did. And revision rewrites the assumption.

Severing "Because" in Real Time

You don't need to wait years to revise. The moment you catch yourself saying "*because*," you can collapse the cause.

- Reaction: "*I'm upset because she ignored me.*"
- Revision: "*That moment is proof that I am always chosen and wanted.*"

The past is not fixed. The script is not final. Revision severs the link and installs a new loop.

The Illusion of Memory

Memory itself is not objective. Two people can remember the same event in completely different ways. What you recall is already filtered by assumption.

When you revise, you're not lying to yourself – you're choosing which version of memory becomes law in your reality.

Identity Statements

- "The past has no power except the story I give it."
- "Revision severs the illusion of cause."
- "I rewrite memory, and reality obeys."
- "Nothing is fixed. My word is final."

Exercise: Collapse a Cause

1. Choose one memory where you've said, "*That event caused my struggle.*" Example: "I failed in business because my first launch flopped."
2. Write the event in neutral terms. Example: "The launch happened."
3. Rewrite the meaning: "*That launch was proof that my success is inevitable.*"
4. Decide this has always been the truth. Live from the new memory.

Closing Reminder

Revision is the knife that severs the illusion of cause and effect. The past is not a chain holding you down. It is clay in your hands.

And when you rewrite it, the reflection changes now.

Integration Notes

"The cause of all results is
who I decide I am."

Chapter 7

Action Is Optional, Assumption Is Law

One of the hardest illusions to break is the belief that action causes results. You've been taught all your life that if you do more, you'll get more. If you push harder, you'll go farther. If you follow the steps, you'll finally succeed.

But action is not the cause. Action is the reflection of the assumption you already hold.

Why Action Feels Like the Cause

Action feels causal because it's visible. You can measure it, point to it, and say, "*This is why I succeeded.*"

But if action were truly the cause, the same action would always produce the same results. It doesn't.

- Two people market their businesses the same way. One thrives, the other fails.

- Two athletes train equally hard. One rises, the other burns out.
- Two writers publish books. One goes unnoticed, the other becomes a bestseller.

If action were the cause, the outcomes would be identical. They're not. Assumption is what makes the difference.

The Real Law: Assumption First

Action flows from identity. When you stabilize an assumption, the actions that match it feel natural and effortless.

- Assume wealth, and you'll find yourself spending, investing, and moving like someone who always has money.
- Assume magnetism, and your body language, tone, and choices shift automatically.
- Assume stability, and your actions carry calm authority.

The action didn't cause the reality. The assumption did. Action was just the echo.

Action Without Assumption

When you act from a shaky or negative assumption, the action is empty.

- Work harder under the assumption "*I never get ahead*," and you'll just reinforce struggle.
- Chase love under the assumption "*I'm not chosen*," and you'll reinforce rejection.
- Push for health under the assumption "*I'm broken*," and you'll reinforce illness.

The action doesn't override the assumption. The assumption always wins.

Why This Frees You

Once you see that action isn't the cause, you're free. You no longer need to grind, hustle, or prove yourself through endless effort.

You can rest in the knowing that assumption is law. Action becomes optional, natural, and effortless. It happens because you are already being the version who has the result.

Identity Statements

- "Action is reflection, not cause."
- "My assumption comes first. My actions follow."
- "The cause of all results is who I decide I am."
- "I never need to hustle. I need only to assume."

Exercise: Trace Your Actions

1. Pick one recent action you took to "make" something happen. Example: sending ten job applications.
2. Write down the assumption underneath it. Example: "*I have to prove myself because I'm not enough.*"
3. Ask: "What would this action look like if I assumed success was already mine?"
4. Reframe the action as reflection, not cause.

Closing Reminder

Action is not law. Assumption is law. When you stop worshipping effort and start anchoring identity, life simplifies.

Because the cause of every result is always assumption → reflection.

Integration Notes

The end comes first, and
the steps arrange
themselves accordingly.

Chapter 8
Collapse the Middle

Most people live trapped in the "middle." They believe life is a sequence of steps: first this happens, then that, and eventually the result arrives.

- "If I work hard, then I'll get the promotion, then I'll make more money."
- "If I go on enough dates, then I'll find the right partner."
- "If I follow the health plan, then I'll finally feel better."

This is the illusion of cause and effect. It keeps you chained to steps, waiting for results that only ever come when you stop waiting.

The Middle Is Made Up

The middle is nothing but filler. It's the sequence you assume is required before you can have what you want.

But reality doesn't need the middle. It only needs the end you choose.

When you collapse the middle, you declare the end now. You don't wait for the promotion, the dates, or the health plan. You step into the identity of someone who already has the money, the love, the health.

And reality reflects it.

Example: Money

- Old story: "I'll save, then I'll invest, then eventually I'll be wealthy."

- Collapsed story: "I am wealthy now. Money multiplies for me easily."

The steps may still appear, but they're no longer conditions. They're reflections.

Example: Relationships

- Old story: "If I heal, then I'll meet someone, then I'll finally be chosen."

- Collapsed story: "I am already chosen. I am always loved."

The dates and people may show up, but they come as reflections of identity – not as prerequisites.

Example: Health

- Old story: "If I rest, then recover, then work out, then I'll be healthy again."
- Collapsed story: "I am healthy now. My body reflects strength and balance."

The steps of rest or recovery may still happen, but they happen inside the story of health already chosen.

Why the Middle Feels Necessary

The middle feels necessary because culture worships process. You've been told your whole life that achievement requires steps.

But steps aren't causes. They're reflections. The end comes first, and the steps arrange themselves accordingly.

Living Without the Middle

When you collapse the middle, life feels simpler, calmer, faster. You don't obsess over "how." You don't chase steps. You only declare the end — and reality rearranges to match it.

Identity Statements

- "The middle is optional. The end is law."
- "I collapse the how and declare the done."
- "Steps are reflections, not requirements."
- "I never wait for the middle. I live from the end."

Exercise: Collapse One Middle

1. Write down something you've been waiting for. Example: "Once I get the promotion, I'll feel secure."
2. Cross out the middle step.
3. Replace it with the end: "*I am secure now. My promotion reflects my security.*"
4. Declare the end daily until the middle loses its grip.

Closing Reminder

The middle is nothing but delay. When you collapse the middle, you stop waiting for steps to justify your truth.

You choose the end, and reality has no choice but to follow.

Integration Notes

"There is no wait. My
reality reflects now."

Chapter 9

Assumption as Immediate Cause

At the core of this book is a radical truth: assumption is not just the first cause – it is the only cause.

There are no chains of events, no hidden forces, no time gaps that create results. There is only the assumption you stabilize and the reflection that follows.

Why Immediate?

Most people accept that "thoughts create reality," but they still believe it takes time. They imagine a lag, a waiting period, a gap between declaring and seeing.

But that lag isn't law. It's only another assumption.

When you decide your assumption is immediate, the reflection appears immediate.

Everyday Proof

Think of moments when you shifted instantly:

- You decided "*This is going to be a good day*," and suddenly everything felt lighter.
- You declared "*I can handle this*," and your body calmed down.
- You assumed "*I'm lucky*," and coincidences started piling up.

Nothing "caused" the shift except the assumption itself. And it showed up right away.

Why People Wait

People wait because they've been taught to. They believe in time as a cause. They assume results must follow later, so they do.

But when you stop assuming delay, you stop experiencing delay.

Assumption as Physics

Your assumptions behave like gravity: invisible, constant, non-negotiable. The moment you stabilize a story, reality bends to it.

- Assume scarcity, and everything reflects lack.
- Assume abundance, and everything reflects overflow.
- Assume love, and everything reflects connection.
- Assume rejection, and everything reflects distance.

There is no wait. The reflection is immediate.

Example: Relationships

- Old belief: "If I work on myself long enough, eventually I'll be chosen."
- Immediate assumption: "I am chosen now. I am already loved."

When you decide this, people in your reality shift immediately. Some lean closer. Some fade away. But the reflection obeys without delay.

Identity Statements

- "Assumption is the only cause, and it is immediate."

- "There is no wait. My reality reflects now."
- "Time is not a law; it's a story. I drop the story."
- "The moment I decide, it is done."

Exercise: Collapse Time

1. Pick one desire you've been waiting for. Example: "I'll be wealthy someday."
2. Rewrite the statement without time. Example: "I am wealthy now."
3. Declare it as immediate law.
4. Throughout the day, notice reflections — small or large — that confirm it now.

Closing Reminder

Assumption doesn't need time. Assumption is law.

The only reason reflection feels delayed is because you assumed delay. Drop that story, and the loop closes instantly.

Integration Notes

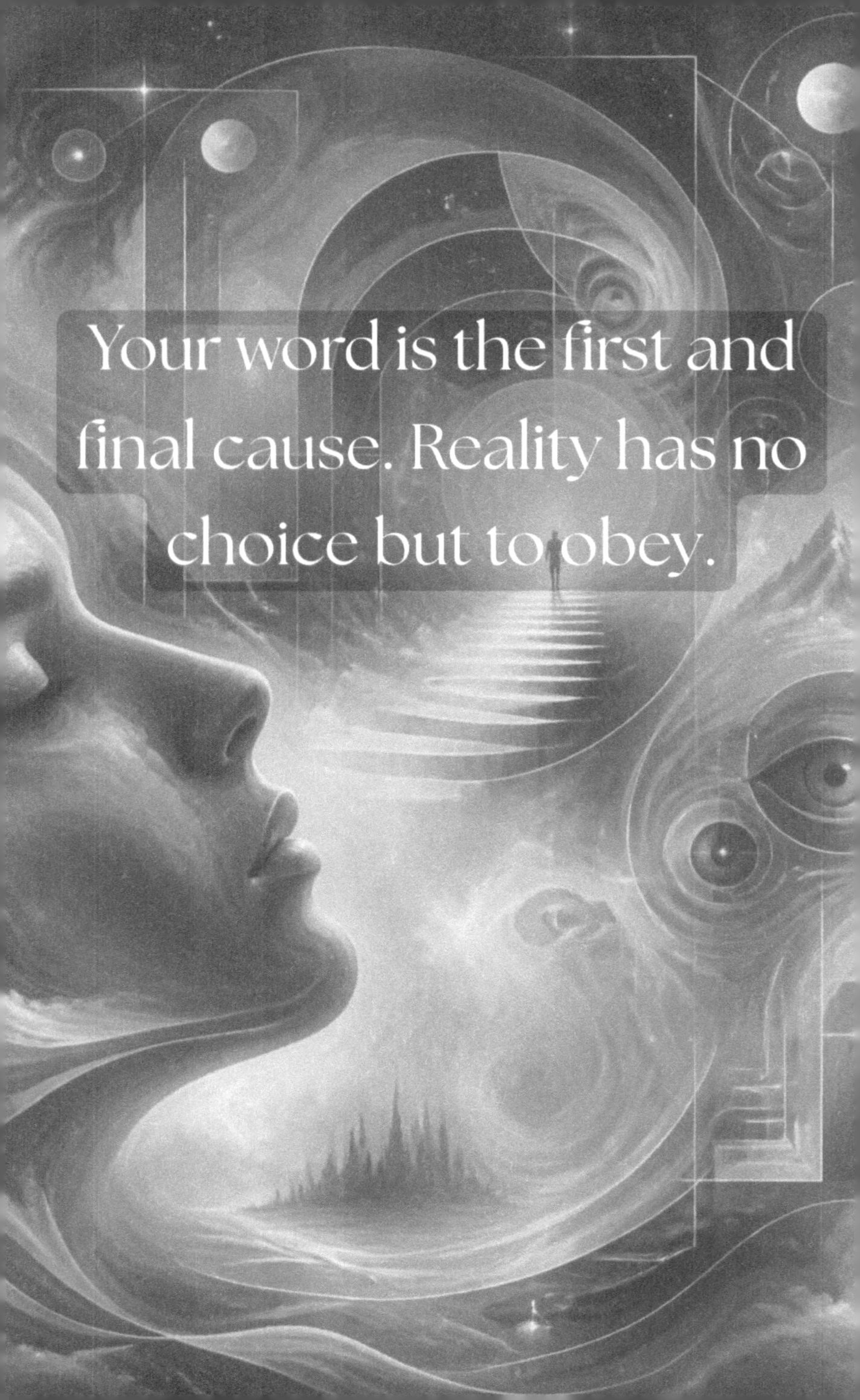
Your word is the first and final cause. Reality has no choice but to obey.

Conclusion: The Freedom of Knowing

Cause and effect is the oldest story humanity tells. It has shaped science, religion, culture, and even personal development. And yet, it is still just that — a story.

When you peel back the illusion, you see the truth that's been hiding in plain sight: Nothing outside of you causes anything. There are no chains, no steps, no delays.

There is only assumption → reflection.

Why This Is Freedom

Once you know this, life becomes lighter. You no longer need to track every event, analyze every step, or obsess over what caused what.

- The past no longer has power, because it was never the cause.
- The present no longer intimidates you, because it's just a mirror.

- The future no longer worries you, because you are the one writing it.

Freedom is knowing the chain was never real.

The End of Blame

Without cause and effect, there's no one left to blame. Not your parents, not your boss, not the government, not even yourself.

Because events don't cause anything. Only assumption does.

And assumption is yours to change at any moment.

Living This Truth

When you drop cause and effect, you stop chasing steps and start stabilizing identity.

- You don't need proof, because reflection is always late.
- You don't need effort, because action is only an echo.
- You don't need to wait, because assumption is immediate.

Your word is the first and final cause. Reality has no choice but to obey.

The Simplicity on the Other Side

What once felt complex becomes simple:

- Decide.
- Know.
- Let reality reflect.

That's it.

No formulas. No chains. No waiting.

Just the freedom of assumption → reflection.

Final Word

Cause and effect is a lie.

And once you stop living inside that illusion, you'll see what was true all along: reality doesn't respond to reasons. It responds to you.

About The Author

The Quantum Shifter is the pen name of an Ivy-trained psychologist, researcher, and lecturer whose work explores the intersection of identity, consciousness, and reality creation. After completing advanced studies in psychology and spending years teaching and writing at the university level, she shifted her focus toward the deeper question that shaped her life's work:

How does identity create reality — and why does changing the story of who you are change the world you live in?

Her publications draw on decades of private research, academic inquiry, and experiential observation. Known for her clarity, precision, and deeply grounded approach, she writes about identity physics, timeline selection, self-concept, and the mechanics of reality in a way that is both accessible and transformative.

Although her academic background is traditional, her pen name reflects the evolution of her work — a nod to the moment she realized that consciousness itself is the most powerful instrument of change. The name stuck, and readers around the world now associate *The Quantum Shifter* with high-level insight, elegant simplicity, and rapid identity transformation.

She currently lives a quiet, contemplative life dedicated to writing, research, and teaching the principles of identity-based reality creation. Her books serve as distilled explorations of the concepts she has spent decades studying — written for readers who are ready to shift who they are, not just what they want.

www.ingramcontent.com/pod-product-compliance
Lightning Source LLC
Chambersburg PA
CBHW031330230926
51848CB00001B/176
* 9 7 8 1 9 7 2 4 3 2 2 5 9 *